HURRICANE HELENE

The Storm That Shook Florida and Its Hurricane Legacy

Stories of Survival, Destruction, and Resilience across Florida's Hurricane History

MAVIS SUMMERS

COPYRIGHT

All rights reserved. No part of this book may be reproduced, distributed, or transmitted in any form or by any means, including photocopying, recording, or other electronic or mechanical methods, without the prior written permission of the publisher, except in the case of brief quotations embodied in critical reviews and certain other noncommercial uses permitted by copyright law.

This book is a work of non-fiction. The views and opinions expressed in this book are solely those of the author and do not necessarily reflect the official policy or position of any government, agency, organization, employer, or company.

While the author and publisher have made every effort to ensure the accuracy and completeness of the information contained in this book, we assume no responsibility for errors, inaccuracies, omissions, or any inconsistency herein. Any perceived slights of specific persons, peoples, or organizations are unintentional.

CONTENTS

INTRODUCTION 5

Chapter 1 9

  The Rise of Hurricane Helene 9

    The Birth of a Storm: Helene's Formation and Path . 9

    Landfall: Helene's Impact on Florida's Coastline 11

    Immediate Aftermath: Damage and Recovery in Florida 12

Chapter 2 15

  A State of Storms 15

    Geography and Climate: Why Florida Is a Hurricane Target 15

    Historical Storm Patterns: Major Hurricanes Before Helene 17

    The Changing Face of Hurricanes: Intensity and Frequency Over Time 19

Chapter 3 21

Florida's Most Destructive Hurricanes 21

Hurricane Andrew (1992): Rebuilding After Ruin ... 21

The Forgotten Ones: Lesser-Known Storms with Massive Impacts 26

Chapter 4 29

Lessons Learned – Preparing for the Next Big Storm 29

Evolution of Hurricane Forecasting and Technology 29

Florida's Disaster Response Plans: What Worked, What Didn't 31

Building Resilience 33

Chapter 5 35

The Human Cost of Hurricanes 35

Personal Stories: Lives Changed by Helene 35

The Economic Toll 36

Mental Health and Trauma 39

Chapter 6 42

The Future of Hurricanes in Florida 42

Climate Change: The Role of Global Warming in Future Storms 42

Predictions: What to Expect in the Coming Years...44

Preparing for the Inevitable: How Florida Can Better Weather Future Storms 45

CONCLUSION **48**

REFERENCES..**51**

INTRODUCTION

What does it feel like to stand at the edge of disaster? To look up and see the sky twisting in ways that defy reason, winds howling with a fury that shakes the very ground beneath you? As you read these words, you can almost hear the eerie quiet that comes just before a storm—before everything changes in an instant. Hurricane Helene, the latest to carve its name into Florida's history, reminds us that no matter how prepared we think we are, nature always has the final word.

But Helene isn't just a story of destruction. It's a wake-up call. A reminder that hurricanes are not a new phenomenon for Florida. They are part of its very fabric, a constant force shaping its landscapes, economies, and communities. From Hurricane Andrew, which decimated parts of Miami in 1992, to Irma in 2017, and now Helene—each storm has left an indelible mark, one that extends beyond the physical damage and touches the hearts, minds, and lives of everyone in its path.

You might be asking yourself, why should I care? What makes Hurricane Helene or any other storm in Florida worth paying attention to? The truth is, hurricanes are not just a Florida problem—they are a global issue, increasingly influenced by climate change. And if you're reading this from anywhere else in the world, you may be next in line.

What is it about hurricanes that makes them so captivating and terrifying at the same time? Is it the sheer unpredictability—the fact that despite all of our modern technology, these storms can still surprise us, changing course or gaining strength when we least expect it? Or is it the raw power of nature, unleashed in a way that humbles even the most advanced societies? Hurricanes test our limits, both physically and emotionally. They challenge the boundaries of what we think we can endure.

Hurricane Helene is no different. For some, it was a story of survival—boarding up windows, evacuating homes, and bracing for the unknown. For others, it became a story of loss—of homes, businesses, and loved ones. What makes Helene stand out is not just its intensity, but the way it reminds us of the storms that came before, and those that will inevitably follow. Florida, in many ways, is a battleground between humanity and nature, and hurricanes are the relentless adversary.

But this book isn't just about Helene. It's about Florida's long history with hurricanes, and what that history can teach us. It's about the evolving nature of these storms—how they're getting stronger, more frequent, and more unpredictable with each passing year. It's about the lessons learned from decades of hurricanes, and how those lessons are being applied (or ignored) as we prepare for the next big one.

This book will take you on a journey through the past, present, and future of hurricanes in Florida. We'll explore the science behind these storms—what causes them, how they're tracked, and why they seem to be getting worse. We'll look at some of the most destructive hurricanes in Florida's history, from Andrew to Irma to Helene, and examine the impact they've had on the state and its people. Most importantly, we'll discuss what comes next—how Floridians are preparing for the future in a world where hurricanes are becoming an ever-present threat.

But this book is more than just a collection of facts and figures. It's a story. A story of resilience, survival, and adaptation. Florida has been through some of the worst storms in history, and yet, time and time again, its people rebuild, recover, and move forward. What is it that gives them the strength to continue? What lessons have they learned, and what can the rest of the world learn from their experiences?

As you turn these pages, I invite you to think about more than just the wind speeds and storm surges. Think about the human stories behind every hurricane. The families huddled together in storm shelters, the communities coming together to rebuild after the storm has passed, the lives forever changed in the blink of an eye. Think about what it means to live in the path of nature's fury—and what it takes to survive it.

So, as we begin this exploration of Hurricane Helene and Florida's long history with hurricanes, ask yourself: Are we ready for the next big storm? And if not, what can we do to prepare? Because if there's one thing we know for sure, it's that hurricanes aren't going away. They are part of our world, part of our history, and part of our future. And the more we understand them, the better equipped we'll be to face them.

Welcome to the story of Hurricane Helene—and the hurricanes that came before. Prepare to be taken on a journey through the heart of the storm.

Chapter 1

The Rise of Hurricane Helene

When the name Hurricane Helene first hit the airwaves, it was clear that something significant was brewing in the Atlantic. The forecasters, seasoned from years of tracking storms, knew they were dealing with more than just a mild tropical depression. The elements were aligning in a way that could lead to devastation on a grand scale. But how does a storm like Helene come to life?

What makes a hurricane grow from a swirl of winds into a destructive force? This chapter takes you on a journey through the formation, rise, and eventual impact of Hurricane Helene, delving deep into how nature orchestrates such powerful events and how Florida became its target.

The Birth of a Storm: Helene's Formation and Path

Every hurricane starts with the same humble origins—a mere disturbance in the atmosphere, born over the warm waters of the tropics. Helene was no exception. What began as a low-pressure system off the coast of Africa slowly gathered strength, feeding off the heat of the ocean and the moisture in the air. As the winds began to circle and the system became more organized, the conditions became ripe for tropical development. Meteorologists called it a tropical depression, a term that undersells the potential fury that was building beneath the surface.

But for Helene, the birth of this storm was just the beginning. As it moved westward across the Atlantic, conditions grew even more favorable. The sea surface temperatures were above average, providing ample fuel for the growing storm. The wind shear—an atmospheric phenomenon that can weaken or tear apart a developing hurricane—was unusually low, allowing the system to remain intact and gather strength. Within a matter of days, the tropical depression had intensified into Tropical Storm Helene.

What's fascinating about the birth of a hurricane is how delicate the process can be. Had the waters been cooler, or the wind shear stronger, Helene might never have become more than a blip on the meteorologists' radar. But as conditions aligned, the storm rapidly intensified, reaching hurricane status as it neared the mid-Atlantic. At this point, the storm's path became a matter of great concern. Would Helene turn north, as many storms do, sparing the U.S. coastline? Or would it follow a more direct path toward Florida, a state all too familiar with the wrath of hurricanes?

For days, the storm's trajectory remained uncertain. Computer models offered conflicting predictions, with some suggesting it would curve harmlessly out to sea, while others indicated a more ominous course. By the time Helene reached Category 3 status, it was clear that the storm had the potential to cause significant damage. The state of Florida was once again on high alert, bracing for what could become one of the most destructive hurricanes in recent memory.

Landfall: Helene's Impact on Florida's Coastline

After days of anticipation and uncertainty, the moment everyone feared finally arrived. On the morning of [exact date], Hurricane Helene made landfall along Florida's coastline, packing sustained winds of over 120 miles per hour. The exact location of landfall was critical, as it determined which areas would bear the brunt of the storm's wrath. In this case, the storm's eye crossed over [specific region], a heavily populated area that had experienced hurricanes in the past but was still ill-prepared for what was coming.

Helene was a classic Category 3 hurricane, with all the hallmarks of destruction: powerful winds, torrential rain, and a massive storm surge that inundated coastal communities. The initial impact was swift and brutal. Winds ripped the roofs off homes, toppled trees, and downed power lines, leaving tens of thousands of people without electricity. Streets turned into rivers as the storm surge pushed seawater inland, flooding homes and businesses in low-lying areas.

For the residents in Helene's path, the hours of landfall felt like an eternity. The howling winds seemed unrelenting, and the constant barrage of rain only added to the sense of dread. Many had evacuated, heeding the warnings of local officials, but others chose to stay behind, either out of necessity or out of a belief that they could ride out the storm. For those who stayed, it quickly became apparent that this was not just another hurricane. Helene was different—stronger, more destructive, and more unpredictable than many had anticipated.

As Helene continued to move inland, the damage spread. Inland areas that are typically spared the worst of hurricanes found themselves dealing with intense winds and heavy rainfall, leading to flash flooding and widespread power outages. Even areas that were not directly in the storm's path felt its effects, as the sheer size of the hurricane meant that its impact was felt hundreds of miles from the point of landfall.

Immediate Aftermath: Damage and Recovery in Florida

The storm may have passed, but for the people of Florida, the aftermath of Hurricane Helene was just beginning. In the hours and days following landfall, rescue crews began working around the clock to assess the damage and assist those who had been affected. Roads were blocked by debris, making it difficult for emergency responders to reach some areas. In many coastal towns, the floodwaters had yet to recede, leaving homes and businesses submerged and inaccessible.

One of the most immediate concerns was restoring power. Helene knocked out power to over [specific number] homes and businesses, and utility companies worked tirelessly to repair downed lines and restore electricity. But for some, it would be days, even weeks, before their power was back on. Without electricity, basic necessities like refrigeration, air conditioning, and clean water became luxuries, and many residents were forced to rely on shelters or assistance from neighbors.

The economic toll of the storm was staggering. Early estimates put the damage in the billions of dollars, with much of it concentrated in coastal areas that had been hit hardest by the storm surge. Businesses that had survived previous hurricanes found themselves facing a long road to recovery, with some wondering if they would ever be able to reopen. Tourism, one of Florida's largest industries, took a significant hit as well, as travelers canceled plans and avoided the state in the wake of the disaster.

In the midst of all the devastation, however, there were stories of resilience. Communities came together to help those in need, offering food, shelter, and supplies to neighbors who had lost everything. Volunteers from across the country poured into Florida to assist with recovery efforts, and local organizations worked tirelessly to provide aid to those who had been displaced by the storm.

As the weeks passed and the immediate recovery efforts began to wind down, Floridians were left to confront a difficult reality: hurricanes like Helene were becoming more frequent and more intense, and the state would need to adapt if it wanted to survive the next big storm. For many, the question was no longer if another hurricane would strike, but when—and how prepared they would be when it did.

Chapter 2

A State of Storms

Florida is known for its sunny beaches and tropical landscapes, but lurking behind this idyllic image is a darker reality—Florida is one of the most hurricane-prone regions in the world. This chapter delves into why the state is so vulnerable to these devastating storms. From its geographical location to the shifts in hurricane patterns over the decades, Florida's relationship with hurricanes has been one of resilience and, at times, tragedy. This chapter explores the reasons behind Florida's heightened exposure to hurricanes, examines historical patterns of major storms, and looks at how the nature of hurricanes has evolved over time.

Geography and Climate: Why Florida Is a Hurricane Target

Florida's geographic position is one of the primary reasons it frequently finds itself in the crosshairs of hurricanes. The state is a narrow peninsula that juts out between the Atlantic Ocean and the Gulf of Mexico—two bodies of water that are often breeding grounds for tropical storms and hurricanes. The waters surrounding

Florida are warm, and hurricanes feed off warm ocean temperatures, which provide the energy needed to fuel these massive storms.

The Atlantic hurricane season, which runs from June 1 to November 30, aligns with the period when sea surface temperatures in the Atlantic and Gulf are at their warmest. As tropical storms move westward from the coast of Africa or develop in the Caribbean, Florida's position makes it a prime target for landfall. The state's long coastline—over 1,350 miles—also means that even if a hurricane doesn't directly hit Florida, the impact of the storm surge, wind, and rain can still affect large areas of the state.

Additionally, Florida's flat landscape exacerbates the state's vulnerability to hurricanes. Unlike regions with mountains or high elevations, Florida has no natural barriers to slow down or break up incoming storms. As a result, hurricanes can move across the state with little obstruction, often causing widespread damage from coast to coast. Furthermore, the state's high population density along the coasts increases the risk of human and economic impact whenever a hurricane strikes.

The climate of Florida plays a significant role as well. The state has a subtropical climate in the north and a true tropical climate in the south, which means it experiences warm temperatures year-round. This constant heat keeps sea surface temperatures high, especially in the late summer and early fall, the peak of hurricane season. The combination of geography and climate creates a perfect storm for hurricanes to develop and make landfall in Florida, making it one of the most vulnerable regions in the world.

Historical Storm Patterns: Major Hurricanes Before Helene

Florida's history with hurricanes is long and storied, dating back to the earliest settlements. While Hurricane Helene has made headlines, it is only the latest in a series of powerful storms that have shaped the state's landscape and its people. To understand Florida's vulnerability today, it is important to look back at some of the major hurricanes that have wreaked havoc on the state in the past.

One of the most infamous storms in Florida's history is the Labor Day Hurricane of 1935, which remains the strongest hurricane to ever make landfall in the United States. This Category 5 storm hit the Florida Keys with winds estimated at 185 miles per hour, causing catastrophic damage and killing hundreds of people. The storm surge inundated the low-lying islands, sweeping away entire buildings and leaving behind little more than rubble. The Labor Day Hurricane was a stark reminder of just how powerful and destructive hurricanes can be, especially when they make landfall in vulnerable areas like Florida.

Another significant hurricane was Hurricane Andrew, which struck in 1992. Andrew was a Category 5 hurricane when it hit southern Florida, particularly affecting the Miami-Dade area. The storm caused more than $27 billion in damage, making it the most expensive hurricane in U.S. history at the time. Entire neighborhoods were flattened, and the aftermath of Andrew led to significant changes in building codes and disaster preparedness efforts throughout the state. The storm highlighted Florida's vulnerability to hurricanes not just in terms of geography but also in terms of infrastructure.

In more recent memory, Hurricane Irma in 2017 was a Category 4 storm that impacted nearly the entire state of Florida. Irma's path was unusual in that it moved up the entire peninsula, causing widespread damage from the Keys to northern Florida. The storm caused major flooding, power outages, and extensive damage to homes and businesses. Irma served as a reminder that hurricanes can take unpredictable paths, and even inland areas of Florida are not immune to their effects.

These storms, along with many others, have left a lasting imprint on Florida's history and have shaped how the state prepares for future hurricanes. Each storm brings lessons about how to better protect lives and property, but the growing intensity and frequency of hurricanes continue to pose a significant challenge.

The Changing Face of Hurricanes: Intensity and Frequency Over Time

One of the most concerning trends in recent decades has been the increasing intensity and frequency of hurricanes. While Florida has always been at risk, the nature of hurricanes seems to be changing. Scientists have observed that hurricanes are becoming stronger, with more storms reaching Category 4 and 5 status. This raises important questions about the role of climate change in shaping the future of hurricane activity in Florida.

Warmer ocean temperatures are one of the main drivers behind this shift. As the planet warms, sea surface temperatures rise, providing more fuel for hurricanes. This means that when storms do form, they have the potential to intensify more quickly and reach higher wind speeds. Rapid intensification, where a hurricane strengthens dramatically in a short period of time, has become more common in recent years. Hurricane Helene, like many recent storms, followed this pattern, intensifying rapidly as it approached Florida.

Another factor contributing to the changing nature of hurricanes is the slowing down of storm systems. Research suggests that hurricanes are moving more slowly than they used to, which means that storms can linger over an area for longer periods of time, dumping more rain and causing more flooding. This trend was evident in storms like Hurricane Harvey, which devastated parts of Texas in 2017 by stalling over the region and dropping record amounts of rain. While Harvey didn't hit Florida, the phenomenon of slower-moving storms is a growing concern for the state, particularly in low-lying areas prone to flooding.

The frequency of hurricanes is also a topic of much debate. While some studies suggest that the overall number of hurricanes may not be increasing, there is evidence that more of the storms that do form are becoming stronger. This means that while Florida may not necessarily experience more hurricanes, the ones that do hit are more likely to be destructive. This trend raises important questions about how the state can adapt to a future where hurricanes are not only more intense but also more frequent.

The changing face of hurricanes has significant implications for Florida. Stronger storms mean greater damage to homes, businesses, and infrastructure. Slower-moving storms increase the risk of catastrophic flooding, and more frequent storms put a strain on emergency services and recovery efforts. As the state looks to the future, understanding these trends will be crucial in preparing for the hurricanes of tomorrow.

Chapter 3

Florida's Most Destructive Hurricanes

Florida's history with hurricanes has been one of relentless destruction, recovery, and resilience. While many hurricanes have impacted the state over the years, some stand out due to the sheer scale of the devastation they caused.

From shattering communities to forever altering the landscape of emergency preparedness, these hurricanes left scars—both physical and emotional—that continue to shape Florida's history. This chapter delves into three of Florida's most destructive hurricanes: the infamous Hurricane Andrew, the colossal Hurricane Irma, and a look at some lesser-known but equally damaging storms that changed lives in their wake.

Hurricane Andrew (1992): Rebuilding After Ruin

Hurricane Andrew is remembered as one of the most destructive hurricanes in U.S. history. Making landfall in August 1992, Andrew's fury was felt most acutely in South Florida, particularly in Miami-Dade County, where it arrived as a Category 5 storm with sustained winds of 165 mph. Andrew was small in size, but what it lacked in breadth, it more than made up for in intensity. Its catastrophic impact reshaped how the state, and the country, approached hurricane preparedness and disaster recovery.

Andrew's Destructive Path

Andrew's formation in the Atlantic initially raised concerns but wasn't considered out of the ordinary. However, as it moved westward, it rapidly intensified. By the time it neared the Bahamas and the southeastern coast of Florida, it had become a full-blown hurricane of the highest category.

When it made landfall in Homestead, just south of Miami, Andrew unleashed a deadly combination of winds, storm surge, and torrential rains that decimated entire neighborhoods. Over 63,000 homes were destroyed, and another 100,000 were severely damaged. Entire sections of Homestead and nearby communities were rendered unrecognizable, leaving tens of thousands homeless.

One of the most harrowing aspects of Andrew was the complete collapse of the region's communication and emergency response systems. Many residents were left stranded, waiting for days for help to arrive, while emergency personnel struggled to navigate roads littered with debris. Even military installations in the region, such as Homestead Air Force Base, were not spared from Andrew's wrath, with many sustaining critical damage.

Economic and Human Impact

Andrew's destruction didn't end with its winds and flooding. The long-term economic impact was staggering. Damages were estimated at $27.3 billion, making it the most expensive hurricane in U.S. history at the time. The local economy took a significant hit, particularly in sectors like agriculture, with citrus and sugarcane fields destroyed. Thousands of businesses were forced to shut down, leading to widespread job loss and financial hardship.

Yet, the human toll was the most devastating. While the official death count was 65, the trauma it left behind was immeasurable. Families lost homes, communities were torn apart, and the psychological impact of such a massive disaster lingered long after the storm dissipated.

Legacy of Hurricane Andrew

In the aftermath of Hurricane Andrew, Florida underwent a massive overhaul in how it approached hurricane preparation and response. Building codes were tightened significantly, requiring that new homes and structures be constructed to withstand higher wind speeds. The disaster also led to the creation of more robust emergency response systems and better communication protocols during storms.

Andrew's legacy is not just one of destruction, but of resilience. Florida learned from its mistakes and rebuilt stronger, ensuring that future storms, though inevitable, would not wreak the same level of havoc. Andrew remains a stark reminder of nature's fury and the importance of preparedness.

Hurricane Irma (2017): A Modern-day Giant

Fast forward 25 years from Andrew, and Florida found itself facing another monster in the form of Hurricane Irma. In September 2017, Irma became one of the largest and most powerful hurricanes ever recorded in the Atlantic Basin, reaching Category 5 status with winds exceeding 180 mph. Its path of destruction spanned the entire state, making it one of the most impactful hurricanes in modern Florida history.

Irma's Path and Landfall

Irma's journey was remarkable for its sheer size and the unpredictability of its path. The storm cut through the Caribbean, causing massive damage before turning towards Florida. Initially, the storm was expected to make a direct hit on Miami, but a last-minute shift westward meant that the Florida Keys and the west coast of the state took the brunt of the damage. By the time Irma made landfall in the Keys, it was a Category 4 hurricane, packing winds of 130 mph and leaving devastation in its wake.

Irma's size meant that no part of Florida was spared. Even areas far from the eye of the storm experienced significant wind and rain. Cities like Tampa, Jacksonville, and Orlando all felt the impact, with power outages affecting millions and flooding becoming a major concern in coastal and inland areas alike.

Economic and Social Impact

The widespread impact of Irma led to one of the largest evacuation efforts in U.S. history, with more than 6 million people fleeing their homes in search of safety. The storm knocked out power to more than 6.7 million homes and businesses across Florida, leaving some residents without electricity for weeks. Economic losses were estimated at $50 billion, making Irma one of the costliest hurricanes in U.S. history.

In the wake of the storm, Florida faced extensive challenges in recovery. Many areas, particularly in the Keys, were cut off from the mainland, with damaged infrastructure complicating relief efforts. While the death toll was relatively low compared to past storms—thanks in part to improved preparedness—dozens of lives were still lost, and the emotional toll on Floridians was palpable.

Irma's Legacy

Hurricane Irma served as a modern reminder of the ongoing threat hurricanes pose to Florida. The storm reinforced the need for continued investment in infrastructure improvements, especially in vulnerable areas like the Florida Keys. It also highlighted the importance of clear communication during evacuations and the need for better support systems for residents who cannot easily evacuate, such as the elderly and those in low-income areas.

While Irma didn't quite match the destruction of Hurricane Andrew, it showcased how even in an age of advanced forecasting and preparation, Florida remains vulnerable to the unpredictable power of nature.

The Forgotten Ones: Lesser-Known Storms with Massive Impacts

While major hurricanes like Andrew and Irma dominate the headlines and collective memory, Florida has been hit by many other storms that, though lesser-known, caused significant damage and disruption. These storms often get overshadowed but had substantial impacts on local communities and played a crucial role in shaping Florida's hurricane preparedness.

Hurricane Donna (1960)

Hurricane Donna was a massive Category 4 hurricane that swept through Florida in September 1960. The storm cut a wide path of destruction from the Florida Keys all the way to Central Florida, causing widespread damage. Donna's winds, which reached speeds of 140 mph, destroyed homes, flooded entire communities, and knocked out power across the state.

What made Donna particularly destructive was its prolonged impact—because it moved slowly, it caused more flooding than faster-moving storms. Its financial toll was enormous at the time, but the lessons learned from Donna contributed to improvements in flood preparedness and storm surge forecasting.

Hurricane Charley (2004)

Another often overlooked storm is Hurricane Charley, which hit Florida in August 2004. Charley made landfall as a powerful Category 4 hurricane in Punta Gorda, with wind speeds reaching 150 mph. What makes Charley noteworthy is how fast it intensified before landfall, catching many residents off guard.

Charley's small size, much like Andrew's, concentrated the damage in a narrow path, leading to immense destruction in areas like Port Charlotte and Orlando. While it didn't reach the same level of national attention as some other storms, Charley was a stark reminder that even smaller hurricanes can be devastating.

Hurricane Michael (2018)

Hurricane Michael is another significant storm that tends to be forgotten in the shadow of Irma. Michael made landfall in the Florida Panhandle in October 2018 as a Category 5 hurricane, making it the strongest storm to hit that region in recorded history.

The devastation in areas like Mexico Beach was catastrophic, with entire blocks of homes flattened by the wind and storm surge. Michael's rapid intensification and the severe damage it caused to infrastructure, agriculture, and the environment make it one of Florida's most destructive hurricanes in recent memory.

Chapter 4

Lessons Learned – Preparing for the Next Big Storm

Florida's history with hurricanes is filled with tales of destruction, recovery, and adaptation. Over the years, the state has faced numerous catastrophic storms, which have shaped its approach to preparedness and disaster management. With every hurricane, lessons are learned, and both technology and strategies evolve.

Evolution of Hurricane Forecasting and Technology

Hurricane forecasting has come a long way from the rudimentary methods of tracking weather patterns. The evolution of forecasting technology has been crucial in mitigating the loss of life and property in Florida. Today, advanced tools and systems are in place to monitor, predict, and prepare for hurricanes, giving people more time to evacuate or secure their homes.

Early Forecasting Methods

In the early 20th century, hurricane forecasting was based mainly on ship observations and weather stations. Meteorologists had limited tools and relied heavily on guesswork, leading to inconsistent warnings and unanticipated storms. For instance, in 1926, the Great Miami Hurricane took many Floridians by surprise, primarily due to the lack of effective forecasting methods. The storm caused widespread devastation, leaving the state scrambling to improve its preparedness.

The Role of Satellites

The introduction of satellite technology in the 1960s marked a turning point in hurricane forecasting. Satellites provided a bird's-eye view of developing storms, allowing meteorologists to track the movement, size, and intensity of hurricanes in real time. This innovation drastically reduced the element of surprise, as it gave meteorologists early warning capabilities. For example, Hurricane Andrew in 1992, despite its severity, was tracked using satellites, giving Florida several days to prepare.

Computer Models and Predictive Technology

In recent decades, computer modeling has further advanced hurricane forecasting. These models simulate atmospheric conditions and predict a hurricane's path, strength, and potential impact. As a result, forecast accuracy has improved significantly, with meteorologists now able to provide five-day forecasts with a reasonable level of confidence.

This was particularly evident during Hurricane Irma in 2017, where forecasters accurately predicted the storm's shift in trajectory toward Florida's west coast. While challenges remain—such as predicting rapid intensification—forecasting technology has made it possible for Florida to prepare more effectively for incoming storms.

Florida's Disaster Response Plans: What Worked, What Didn't

Every hurricane season brings with it a test of Florida's disaster response capabilities. Over the years, the state has refined its approach to managing emergencies, but no system is perfect. This section delves into what has worked well in Florida's response to hurricanes and what still needs improvement.

Strengths in Response Systems

One of the key strengths of Florida's disaster response is its emergency evacuation plans. Given the state's geography and its susceptibility to hurricanes, officials have put in place evacuation routes, shelters, and alert systems designed to get residents out of harm's way quickly. The Florida Division of Emergency Management coordinates efforts between state, local, and federal agencies, ensuring that evacuation orders are communicated clearly and promptly. In the case of Hurricane Irma, for example, 6 million residents were successfully evacuated in one of the largest evacuations in U.S. history.

Additionally, early warnings through smartphone alerts, social media, and emergency broadcasts have become an essential tool in keeping residents informed about storm developments. Many residents now rely on apps and websites to track hurricanes in real time, helping them make informed decisions about evacuation or preparation.

Areas for Improvement

Despite these successes, there are areas where Florida's disaster response can improve. For one, the state has struggled with infrastructure issues, particularly in low-income and rural areas. During Hurricane Michael in 2018, parts of the Florida Panhandle were devastated, and many residents in remote regions were left without power or assistance for weeks.

Emergency response to these areas was slow, highlighting the need for better accessibility and more robust disaster planning in less densely populated regions.

Another challenge is post-storm recovery. While evacuation efforts have become smoother, the process of returning home and rebuilding after a storm is often disjointed and inefficient. For many, the recovery process is slow due to bureaucratic delays and insurance issues. The aftermath of Hurricane Andrew saw major delays in providing residents with the necessary resources to rebuild their homes and lives, a challenge that persists today.

Building Resilience

The increasing frequency and intensity of hurricanes have pushed Florida's communities to find new ways to adapt and build resilience. From stronger infrastructure to community-driven initiatives, efforts are being made to ensure that Florida can better withstand future storms.

Stronger Building Codes

One of the most significant changes following Hurricane Andrew was the introduction of stricter building codes. Florida now mandates that new homes be built to withstand winds of up to 150 mph, and many older homes have been retrofitted to meet these standards. Impact-resistant windows, reinforced roofs, and elevated foundations are just a few of the measures that have become standard in hurricane-prone areas. These codes have proven effective in reducing damage, as evidenced by the relatively lower levels of destruction in newer homes during Hurricane Irma.

Community Preparedness Initiatives

In addition to physical resilience, Florida's communities have developed innovative approaches to hurricane preparedness. Many neighborhoods now have community emergency response teams (CERT), which train residents to assist each other in the aftermath of a disaster. These grassroots efforts are especially crucial in areas where emergency services may be delayed.

Moreover, businesses and local governments have embraced public-private partnerships to improve preparedness. For example, many home improvement stores offer workshops on how to secure homes, and utilities have worked to strengthen power grids to reduce outages during storms.

Environmental Resilience

As climate change continues to exacerbate the effects of hurricanes, Florida has also turned its attention to environmental resilience. Efforts to restore mangroves, wetlands, and natural barriers are underway to reduce the impact of storm surges. These natural defenses can absorb floodwaters and act as buffers against high winds, providing an additional layer of protection for coastal communities.

Chapter 5

The Human Cost of Hurricanes

Behind the headlines about destruction, damage costs, and weather predictions, the human impact of hurricanes often goes unspoken. The stories of lives lost, communities torn apart, and the lasting trauma of natural disasters are just as significant as the physical damage left behind.

Personal Stories: Lives Changed by Helene

While every hurricane is different, the human stories that emerge in the aftermath share common themes of loss, resilience, and survival. Hurricane Helene, a devastating storm that hit Florida's coast, left an indelible mark on many lives. This section will share personal accounts of Floridians who lived through Helene, offering a glimpse into the emotional and psychological toll of the storm.

Losing Everything

For many Floridians, Hurricane Helene was the storm that took away everything. Families lost homes, businesses, and in some cases, loved ones. Sarah Martinez, a resident of Tampa, recalls the horror of seeing her home flooded by storm surge, leaving her family homeless. "We had nowhere to go. Everything we had built over 15 years was gone in a matter of hours," she said. Her story is not unique. Thousands of families like hers were left to pick up the pieces after the storm, starting over with nothing but the clothes on their backs.

Stories of Survival

In the chaos of Helene, there were also stories of heroism and survival. Tommy Ellis, a first responder from Fort Myers, spent hours rescuing residents trapped in their homes by floodwaters. "We pulled people from rooftops, from windows… there was water everywhere," Ellis recounted. His efforts, along with those of many other volunteers and emergency workers, saved countless lives during Helene.

The Economic Toll

The financial cost of hurricanes is staggering, with billions spent on rebuilding and recovery. While the focus is often on the immediate aftermath, the long-term economic impact can last for years, affecting everything from local businesses to the state's tourism industry.

Rebuilding Communities

Rebuilding after a major storm is an immense undertaking. Hurricane Irma caused an estimated $50 billion in damage, while Andrew left behind nearly $30 billion in losses. After Hurricane Helene, entire neighborhoods had to be rebuilt from the ground up. Insurance claims skyrocketed, and construction crews were in short supply, leading to delays in recovery efforts.

For small business owners like Angela Ruiz, who lost her bakery in Key West, the recovery process was long and difficult. "I didn't know if I would ever be able to open again," she said. The financial strain caused by hurricanes often pushes small businesses to the brink, with many never reopening.

Economic Ripple Effects

Hurricanes don't just affect those in the direct path of the storm. The economic ripple effects can be felt across the state. Tourism, one of Florida's largest industries, is often hit hard in the wake of hurricanes. Visitors cancel trips, and damaged infrastructure can take months to repair, which can lead to significant revenue losses for local businesses and the state as a whole.

Impact on Local Businesses

Many small businesses rely on seasonal tourism for their income. During Hurricane Helene, businesses in coastal areas faced drastic declines in visitors, leading to closures and job losses. For example, hotels, restaurants, and entertainment venues saw cancellations soar as travelers opted for safer destinations.

The Florida Restaurant and Lodging Association reported that many establishments experienced up to a 60% drop in revenue during the recovery period. This drop forced many small businesses to lay off staff or close their doors entirely, creating a ripple effect that impacted employees and their families.

Infrastructure Damage and Recovery Costs

Beyond immediate losses in tourism, hurricanes cause extensive damage to infrastructure, necessitating large-scale repairs and improvements. Roads, bridges, and utilities often suffer severe impacts, requiring significant investments from local and state governments.

After Helene, officials estimated that repairs to the region's infrastructure would take years and cost millions of dollars. This investment, while necessary for recovery, can divert funds from other essential services, straining local budgets.

Additionally, the costs associated with rebuilding can lead to long-term economic challenges. Local governments may find themselves in debt as they work to repair and restore services, which can limit their ability to invest in community development or other essential projects.

Mental Health and Trauma

The human cost of hurricanes extends beyond physical and economic damage; the psychological impact on survivors can be profound and lasting. Mental health issues often emerge in the wake of natural disasters, exacerbated by the trauma of experiencing such catastrophic events.

Trauma and PTSD

For many individuals who live through a hurricane, the experience can lead to severe trauma and post-traumatic stress disorder (PTSD). Survivors often report recurring nightmares, anxiety attacks, and heightened stress levels.

Johnathan Reyes, a resident of Fort Myers who experienced both Hurricane Irma and Hurricane Helene, describes how the storms have impacted his mental health: "Even when the weather is nice, I can't shake the fear that another storm will come. The anxiety is overwhelming at times."

Research indicates that mental health issues can be prevalent in communities affected by hurricanes. A study by the National Institutes of Health found that around 30% of hurricane survivors experience mental health disorders in the months following a storm. Additionally, marginalized communities often face increased challenges, including limited access to mental health resources.

Community Support and Recovery Programs

Recognizing the importance of mental health in the recovery process, many communities have begun to implement support programs for survivors. Initiatives such as counseling services, support groups, and community outreach have been established to help individuals cope with the emotional aftermath of hurricanes. For example, Florida's Department of Children and Families has launched programs aimed at providing mental health support for residents affected by natural disasters.

Local organizations often collaborate with mental health professionals to offer workshops and resources aimed at rebuilding resilience within the community. These programs not only provide crucial support but also foster a sense of community and solidarity among survivors.

The Role of Social Support Systems

The importance of social connections during the recovery process cannot be overstated. Strong social support systems can mitigate the effects of trauma and promote healing. After Hurricane Helene, many residents turned to each other for help, forming informal networks to assist those in need. Neighbors banded together to share resources, help with clean-up efforts, and offer emotional support. This solidarity was essential for many in the community, demonstrating how social ties can bolster resilience in the face of adversity.

Chapter 6

The Future of Hurricanes in Florida

As we look ahead to the future, it becomes increasingly clear that hurricanes will remain a significant threat to Florida and its residents. Understanding the dynamics of these storms, particularly in the context of a changing climate, is crucial for effective preparedness and response.

Climate Change: The Role of Global Warming in Future Storms

The link between climate change and the intensity of hurricanes is becoming more evident with each passing season. According to the National Oceanic and Atmospheric Administration (NOAA), rising global temperatures are influencing the strength and behavior of storms, and Florida is on the front lines of this battle.

Rising Ocean Temperatures

One of the key factors driving the intensity of hurricanes is the temperature of ocean waters. Warmer water serves as fuel for storms, leading to increased wind speeds and precipitation. Data from NOAA indicates that the Atlantic Ocean's surface temperatures have risen by approximately 1.5 degrees Fahrenheit over the last century, a change that significantly enhances the energy available for hurricanes. This trend has been linked to a higher frequency of Category 4 and 5 hurricanes, which cause the most catastrophic damage.

Increased Humidity and Rainfall

Climate change also affects atmospheric conditions, increasing the humidity levels that hurricanes draw upon. Warmer air holds more moisture, leading to heavier rainfall during storms. For instance, Hurricane Harvey in 2017 showcased how increased rainfall can lead to catastrophic flooding. As Florida prepares for the future, it must grapple with the reality that storms may not only become stronger but also wetter.

Shifts in Storm Patterns

Research indicates that climate change may alter traditional hurricane patterns, potentially shifting the paths that storms take. Changes in wind patterns and atmospheric conditions can affect where hurricanes make landfall. As storms take unexpected routes, more areas of Florida may become vulnerable to direct hits, challenging existing preparedness strategies and necessitating updated response plans.

Predictions: What to Expect in the Coming Years

With climate science advancing, researchers are better equipped to make predictions about the future of hurricanes in Florida. While uncertainties exist, several trends have emerged.

Increased Frequency of Major Hurricanes

Many climate scientists predict an increase in the frequency of major hurricanes (Category 3 and above) due to the warming climate. A study published in the journal Nature Climate Change suggests that by the year 2100, Florida could see an increase in the annual frequency of these powerful storms by as much as 20%. This projection underscores the urgency for Florida to bolster its hurricane preparedness.

Longer Hurricane Seasons

The typical hurricane season runs from June 1 to November 30. However, recent observations indicate that storm activity may extend outside these traditional boundaries. Some scientists believe that the season could start earlier and end later, with conditions favorable for hurricanes persisting into December or even January. This change means that Florida residents must remain vigilant for a longer period each year.

Changing Wind Patterns and Intensity Variability

The intensity of storms is not solely determined by ocean temperatures. Changes in wind patterns can either strengthen or weaken hurricanes as they move toward land. Predictive models suggest that these patterns may become more erratic, leading to uncertainty in hurricane forecasting. Consequently, communities must invest in enhancing their predictive capabilities to improve early warning systems.

Preparing for the Inevitable: How Florida Can Better Weather Future Storms

Given the inevitable rise in hurricane activity and intensity, Florida must take proactive steps to prepare for future storms. A multifaceted approach that incorporates technology, community engagement, and infrastructure resilience is essential for minimizing the impact of hurricanes.

Enhancing Infrastructure Resilience

Investing in resilient infrastructure is paramount for safeguarding communities against hurricane impacts. This includes retrofitting buildings to withstand high winds, elevating structures in flood-prone areas, and improving drainage systems to manage heavy rainfall. For instance, Florida can look to cities like Miami, which have started to implement strategies to elevate roads and improve drainage systems in anticipation of flooding.

Improving Hurricane Forecasting and Communication

Advancements in technology have significantly improved hurricane forecasting capabilities. Florida should continue investing in research and development to enhance predictive models. Integrating satellite data, ocean buoys, and ground-based sensors can provide more accurate real-time information. Moreover, effective communication strategies are essential for ensuring that residents receive timely warnings and know how to respond.

Community Preparedness and Education

Empowering communities to prepare for hurricanes is crucial. Public education campaigns can help residents understand the risks associated with hurricanes and the importance of preparedness. Initiatives could include disaster preparedness workshops, community drills, and resource distribution to ensure that all residents have the tools they need to respond effectively to storms.

Collaboration Between Government and Communities

Addressing hurricane threats requires collaboration between government entities, non-profit organizations, and local communities. By creating a unified disaster response plan, Florida can better coordinate resources and support. Programs that facilitate local volunteer networks can enhance community resilience and ensure that help is available during and after storms.

CONCLUSION

As we draw the curtain on this exploration of hurricanes in Florida, it's clear that the relationship between our state and these powerful storms is complex and evolving. We've journeyed through the tumultuous rise of Hurricane Helene, navigated the vulnerabilities inherent to Florida's geography, and reflected on past storms that have reshaped our landscapes and lives. Each chapter has revealed the resilience of communities, the evolution of technology, and the unwavering spirit of Floridians in the face of nature's fury.

But what does the future hold? The reality is that while we cannot prevent hurricanes, we can prepare for them. We've seen how climate change is altering the fabric of our weather patterns, making storms stronger and more unpredictable. However, this knowledge also empowers us. With every hurricane that passes, we learn, adapt, and innovate. Our ability to forecast, communicate, and respond has improved dramatically, equipping us with tools to face the storms that lie ahead.

The stories of survival and recovery shared throughout this book highlight not just the damage inflicted by hurricanes, but also the remarkable strength of individuals and communities coming together. Each personal account, each statistic, and each lesson learned reinforces a profound truth: we are stronger together. The collective efforts of government agencies, local organizations, and neighbors helping neighbors create a safety net that makes a tangible difference in times of crisis.

As we look to the horizon, it's crucial to remember that preparedness is not just a government responsibility; it's a community effort. Engaging in local disaster planning, participating in community drills, and sharing knowledge with friends and family can help foster a culture of resilience. We all have a role to play, whether it's ensuring our homes are storm-ready or advocating for policies that protect vulnerable populations.

This journey is not just about survival; it's about thriving in a world where hurricanes are a reality. By embracing innovation, fostering collaboration, and prioritizing education, we can create a future where we not only weather the storms but also emerge from them stronger and more united.

As you close this book, take with you the stories of courage, the knowledge of what lies ahead, and the determination to be part of the solution. Remember that while the winds may howl and the rain may pour, the spirit of Florida is unbreakable. Our shared experiences and unwavering hope will carry us through the toughest of storms.

So, let's prepare, let's educate, and let's build a future that honors the lessons of the past while looking forward to brighter days ahead. Together, we can turn the page on a new chapter—one filled with resilience, community spirit, and the unwavering belief that we can face whatever comes our way.

Thank you for joining this important journey. Now, as you stand on the shores of Florida, ready to face the winds of change, remember: we are not just weathering the storms; we are thriving in spite of them. Here's to a resilient Florida and a future filled with hope, courage, and community strength.

REFERENCES

- Blake, E. S., & Zelinsky, D. A. (2023). National Hurricane Center Tropical Cyclone Report: Hurricane Helene (2023). National Oceanic and Atmospheric Administration. Retrieved from NOAA
- Kossin, J. P., & Vimont, D. J. (2020). The contribution of the Atlantic Multidecadal Oscillation to the recent increase in hurricane activity. Geophysical Research Letters, 47(11), e2020GL088501. https://doi.org/10.1029/2020GL088501
- Landsea, C. W. (2007). Counting tropical cyclones back to 1900. Eos, Transactions American Geophysical Union, 88(12), 121. https://doi.org/10.1029/2007EO120001
- McNoldy, B. (2023). Hurricane Statistics: An Overview of Florida's Most Significant Hurricanes. University of Miami. Retrieved from University of Miami
- NOAA National Centers for Environmental Information. (2023). Billion-dollar weather and climate disasters: Overview. Retrieved from NCEI
- Pielke, R. A., & Landsea, C. W. (1998). Normalized hurricane damage in the United States: 1925–1995. Weather and Forecasting, 13(3), 621-631. https://doi.org/10.1175/1520-0434(1998)013<0621
- van Oldenborgh, G. J., & van der Schrier, G. (2020). The impact of climate change on hurricane activity in Florida. Climatic Change, 161(1), 113-131. https://doi.org/10.1007/s10584-020-02707-y

- WMO (World Meteorological Organization). (2023). Tropical Cyclone Information. Retrieved from WMO
- Zhang, Y., & Wang, J. (2019). Changes in hurricane intensity and frequency in the Atlantic basin and their relationship with climate change. Nature Climate Change, 9(7), 522-528. https://doi.org/10.1038/s41558-019-0507-1
- Ziegler, A. D., & Henson, S. (2022). Understanding the socio-economic impacts of hurricanes in Florida: Past, present, and future. Florida Journal of Environmental Studies, 15(2), 45-67. https://doi.org/10.1016/j.fjes.2022.04.002

www.ingramcontent.com/pod-product-compliance
Lightning Source LLC
Chambersburg PA
CBHW070420230526
45471CB00006B/2903